F. T. Piggott

The Garden of Japan

A Year's Diary of its Flowers

F. T. Piggott

The Garden of Japan
A Year's Diary of its Flowers

ISBN/EAN: 9783337124069

Printed in Europe, USA, Canada, Australia, Japan

Cover: Foto ©berggeist007 / pixelio.de

More available books at **www.hansebooks.com**

The
Garden
of Japan.

THE GARDEN

OF JAPAN.

A YEAR'S DIARY OF ITS FLOWERS

By

F. T. PIGGOTT.

WITH FOUR PICTURES BY

ALFRED EAST, R.I.,

GEORGE ALLEN,
RUSKIN HOUSE, LONDON.

SECOND EDITION,
1896.

To all who have wandered in The Garden.

I have endeavoured to keep the Illustrations of the Book at a very respectful distance from the Pictures with which Mr. Alfred East has so kindly embellished it.

The formal decorative designs used on the Title-page and elsewhere need some explanation. They are tracings from the gold ornamentation on lacquer boxes—black having been used in the place of the gold. The design on the Dedication page is of "The Three Friends in Winter," alluded to in the text.

The end-papers have been reproduced from the design on a lady's kimono. It represents the rafts of flowers which are sent floating down the river at the festival in the month of August, held in honour of the Goddess of Broideresses.

The cherry design used for the border of the cover is from a surimono, by Keisai Yeisen.

The little flower-types, used to divide the entries of the Diary, have been specially cast by Messrs. Clowes, from Japanese models.

The Diary was written in the year 1890.

F. T. P.

.

Edgeworthia papyfera
Matsumura   Hearii

Prunus Communis
Summer
April

Prunus Persica
Momo
Early April

Prunus
Kuru
March

Illicium Religiosum
Shikimi
March

Himarichia Japonica
Momiji
Early March

Prunus Ind...
Yma Momé
Feb - march

Prunus Mumé
Momé
March

Prunus mume...
Momi-Mumé   Late Mume

January.

January 1.—Half the entire population of Japan, grown and ungrown, is at the present moment, the afternoon of New Year's Day, flying kites or playing battledore and shuttlecock —the kites making a musical murmur in the air with their cane vibrators; the bats gorgeous with historical romances brocaded and papered into high relief—the other half is paying New Year's visits *pour féliciter*, as the French say. We also write "*p. f.*" upon our cards, and assist in the holocaust of pasteboard which is going on. Also there is much wearing-out of many gorgeous uniforms, for these visits are paid *en grande tenue*, and the streets simply glisten with gold lace. It is indeed the time of delights. The procession of cocked-hats passes under an arcade of crimson and white. At the doorway of every house in the land, from the almost-untidy hovel to the palace, wave two national flags placed crosswise—the red ball for the sun upon a white field of morning light—and underneath the flags a trophy of lobster, seaweed, straw rope, and paper *gohei*; emblems of happiness and long life. Before each door, too, are planted Pine branches and clumps of waving Bamboo, emblems also, of passing time.

> Kadomatsu wa
> Meido no tabi no
> Ichi-ri-zuka
> Medetō mo ari
> Medetō mo nashi.

So sang the Abbot Ikyu Osho of Kyoto long ago of the New

E

Year's festival, and his song may be translated somewhat as
follows :—

> *I must remember sad days ;*
> *But thou, think thou of glad days*
> *While the pine-trees standing great*
> *New Year's pilgrims in the street,*
> *Green milestones as they travel,*
> *Some weeping, some with laughter,*
> *From this life to unravel*
> *The calm that follows after.*

Nature is somewhat sombre at this time ; she has more than
enough to do to keep her offspring alive through the cruel
piercing winds that sweep over the country, following the
unnatural heat of the midday sun as surely as night follows
day ; but she does her best to add to the pleasantness of things
in general. If Pine branches are somewhat stiff and formal, the
Bamboo waves and bends its jointed stems whichever way the
wind may blow, whether it rushes straight down the street, or
seems, where the crossways meet, to come from every quarter
under heaven. And there are plenty of small Jonquils (*Narcissus
Tazetta v. chinensis*) for the vases ; and the crimson-berried
Nanten (*Nandina domestica*) for the pots, whose clusters of shining
red serve us well for Holly berries, with its scarcer yellow sister
for variety. We adapt ourselves to many far-eastern modern
ways, but still we do our best to preserve childhood's traditions
of the home four thousand and odd miles away : so the *Nanten*
is an especial favourite just now.

A pink-berried Mistletoe (*Loranthus Yadoriki*, or a species
of *Viscum* distinct from *V. album*) helps to keep up the illusion
of an absent yet present Yuletide—somewhat, but not quite

2

successfully ; for as some old singer said, "the white is better than the red"—better, I suppose, for those quaint youthful acts of ours on which we were wont to invoke its silent benison. Some few homely memories this strange country preserves for us by means of its flora, in which, as the learned botanists tell us, Nature, by some quaint trick of her own invention, has contrived to mingle the loveliness of the Alps with the splendour of the Tropics. Even Nature has been eclectic : no wonder therefore that man has, in all Japanese time, followed suit. But she does not allow us to forget that she is Japanese here, and so she gives us something quite unique in the tiny plum-trees (*Prunus mume*), with the tiny buds which are just opening into flowers.

The Pine, the Bamboo, and the Plum are the "Three friends in winter"—*Saikan no san yu*—and they are used as the bearers of good wishes for the New Year: the Pine for longevity, the Bamboo for uprightness, the Plum for sweetness.

To the Japanese people life would indeed be dreary and monotonous without the flowers, and the desire for them is gratified all the year round, now with profusion, now scantily ; but still there is always something, some spot of Nature's colouring, on which the eye may gladly rest during those odd moments of leisure which so abound in the Japanese day. Such genuine, and withal such peaceful admiration I have never seen before, and probably shall never see again after I leave this land of ever-opening flowers. In what it differs from our western ways of admiring it is difficult to say. We are not unconscious of the softening influences of Nature's choicest works. Herrick has sung of their beauties, and of the lessons of their little lives, as no Japanese poet even has ever sung—though somehow we seem to take it all for granted ; but in Japan the flower, and more especially the first of its kind, is actually greeted with a solemn and becoming reverence. The admiration is active, not passive ; it is derived, not from a passing glance, but from a steady lingering look ending in one of those soft gurgling

3

sounds which are potential in expression of infinite and complete satisfaction.

Looking back through the year just past, the path of Time seems literally to have been strewn with bright falling petals. It matters little what the incident may be on which one's thoughts linger, it had always its floral accompaniment; now rich in tone and colour, now sad and plaintive, as the seasons passed. So pleasant has this made the memories of the days, that this year I have resolved to keep a floral diary, jotting down therein the story and the songs of the flowers by the wayside, as they bud and bloom and fade. It will, I fear, be unscientific to a degree: nothing but casual notes and recollections, very far from complete; the slightest of records; faint impressions of seasons that I would were both of them not quite so fleet of foot: of the weeks and days of winter growing gradually in delight as they brighten into spring: of the springtime as it blossoms into summer: of the over-heated summer as it mellows into the full rich-tinted autumn, till that too fades and dies, and sharp winter comes again. Maybe its only value will be to recall to some who have wandered there one that is not the least of the charms of this beautiful land.

To the short catalogue of flowers noted on this first of January, I have to add a bright yellow *Arabis*.

I have also a stray branch of single crimson Camellia gathered in one of the sheltered valleys which abound in Tokyo; but the time of the Camellias is not yet, a few buds only have been deceived by the noontide brilliancy of the sunshine into premature flowers. The garden is brown and bare and uninteresting as a door-mat, and will scarcely repay a day's labour, though it costs but thirty cents. There is, it is true, much promise for the future; the lingering warmth of the late autumn days pushed on apace the bud preparation; but we will not disturb them for the present, rather let them sleep on in their

4

cases. The trees are bare, the turf to all appearance dead and past regeneration, at least to those who do not know the habits of Corean grass; the sub-tropical plants, the Palms and Bananas, are wrapt up in their warm winter dresses (literally "dresses"— *kimono* in the Japanese). These are made of straw, of course— everything is made of straw that is not made of bamboo; but, of course, in the hands of the Japanese they become quite graceful.

A SAGO-PALM IN ITS WINTER DRESS.

It is at this time of the year, however, that the floral instinct of the people is especially noticeable. None of the many sights in the streets which strike strange eyes as eccentric, is so strange as to see people carrying home, with a tender care bred of admiration, big bunches of bare twigs, with perhaps not more than two or three half-open blossoms. After a while one gets accustomed to it, as one gets accustomed indeed to other amiable eccentricities; but in the fulness of time we come to learn what the flower-worship of the Japanese really is, and to realise that it is a thing which has not been, as yet, vouchsafed to

5

our civilisation. We come to learn that flowers and leaves are not the only parts of Nature's handiwork worth setting store by ; that they are but part of the system which includes twigs and branches as well ; and that it is in the completed system of flower and leaf and bough that our great art-mistress has elaborated the fundamental laws of decoration. At length we come to realise that the pleasure in the whole must be greater than the pleasure in its part. The science which the Japanese have constructed for the arrangement of flowers, based as it is on a thorough knowledge and appreciation of the laws of beauty and harmony, is one of the most striking evidences of the perfection of their civilisation. The whole scheme of the science has been mapped out for us by a patient and enthusiastic student, and does more than well repay perusal. Here I need only dwell for a moment on the pleasure-giving capacity of these bare twigs, which so abound just now. The branch, set in due order in its vase, the beauties of its lines emphasised, delights us first ; and then day by day it gives evidence of its vitality, as one by one the budlets, of which at first we were unconscious, grow to buds, then blossom into flowers, until the whole branch is covered as it were with a snow-white veil. Thus, after the pleasure in the arrangement, comes the pleasure in the flowers themselves ; and to this is added the true joy of the gardener, which delights in watching the gradual growth and development of his treasures.

I should trespass too much on ground which belongs entirely to Mr. Josiah Conder, if I dwelt at any length on the beautiful science of Japanese flower arrangements. I must content myself with giving here a typical example of the arrangement of a bough of plum-blossom, and occasionally a few other examples reproduced from the native works on the subject. Mr. Conder's delightful book, " The Flowers of Japan, and the Art of Floral Arrangement," is full of such drawings, and sets out for us the whole of the art and the mystery.

On the other hand, however, there is a mystery about this

6

ARRANGEMENT OF BOUGH OF PLUM-TREE.

matter which I confess I am quite unable to solve. The demand for these branches is inexhaustible, and so apparently is the supply ; and yet it takes the growth of many years to furnish the adornment of as many days. This reckless and ruthless lopping of big branches jars somewhat upon my gardening instincts.

January 7.—To-day is the domestic festival of the "seven herbs"—*Nana kusa*—one of those old-time ceremonies which abounded in Japan, but which are fast passing out of memory. The master of the house, rising betimes, and robed appropriately, proceeded to the kitchen ; and there, in the presence of his servants, he took an equal quantity of each of seven herbs, which he chopped up fine. The herbs were then boiled in a species of rice gruel, and served for breakfast to all the household, with an ordered ceremonial that only the Japanese mind could conceive.

The names of the herbs form a verse of 5, 7, 5, 7, 7 Japanese poetry :—

> Seri, Nadzuna,
> Gogiō, Hakobera,
> Hotoke-no-za,
> Suzuna, Suzushiro,
> Korede Nanakusa,

which being translated botanically runs thus :—

> *Œnanthe stolonifera, Capsella Bursa-Pastoris,*
> *Cratæva falcata (?), Stellaria media,*
> *Lamium amplexicaule,*
> *Brassica campestris, Arabis flagellosa,*
> *These are the " Seven-herbs."*

February.

February 6th.—A long interval since my last entry, which has been filled almost exclusively with plum-trees and plum-branches. In the garden nothing has occurred to vary the monotony of Nature's most uninteresting proceedings — I should say, absence of proceedings. This long sleep makes one think that, perhaps after all, the flowers of Japan may only be the fairy flora of a dream.

The cold dry weather has given place to snow, a variation which is not specially charming, but the farmers are pleased, so there is nothing more to be said ; after all, snow, if only it would not turn quite so rapidly to slush, is neither more nor less unpleasant than driving north-east winds and dust. The *status quo* of impatient expectancy is maintained. To-day, however, I can make an addition to my still slender catalogue ; some little yellow flowers, on the barest of bare twigs of course, force themselves on our notice, in spite of, and through the snow which almost hides them. They are welcome little strangers, to me quite unknown. The coolies insist on calling the trees yellow Plums ; but my more accurate gardener gives them a name which I have since been able to convert into *Chimonanthus fragrans*—japonicé *Nankin-mume*—the "Nankin plum." The flower is insignificant, and does not cluster about the stems so freely as the pink or the white plum ; but it is curious, and

c

makes a pleasant little addition to our stock of colour. A further note on this flower will be found in April.

Now, as for these Japanese names, I think they had better be omitted in future, except where they have passed into our botanical vocabulary. Mr. Matsumura's valuable catalogue of the flowers of Japan enables me to trace the classical name after I have, not without difficulty, deciphered my gardener's coolie pronunciation of the native name.

Latter End of February.—Old winter lies a-dying; and these many days past we have been busy gathering, not rosebuds, but every other sort of bud while we might; "buds which," as the Japanese saying is, "will never see the leaves." Like King Charles' head of undying memory, I find these buds cannot be kept out of my diary for the present. They are the only food on which our æsthetic feelings are fed now, and so we continue breaking off big boughs in reckless fashion: we continue tenderly to carry them home, and—having set them in the most honourable place—to watch, with ever-growing delight, the old round of bud and blossom and falling petals. The process of development is a little more rapid now, for the midday warmth quickens the rising sap. Again I wonder, as I see the flower-shops and barrows filled with branches, whether the lucrative business engendered by the bud craze is not too practical an example of the bird in the hand. Surely no garden can stand it! and then there will be no flowers in the bush! I reflect sadly that it is all too characteristic of the people to whom the pleasure of to-day is all in all; these light-hearted folk always let to-morrow look after itself. If there should be no flowers, well *shikata ga nai*—it can't be helped—we must try and grow some for next year. In *shikata ga nai* the whole philosophy of the nation lies embedded; paraphrased, it seems to be equivalent to "eat, drink, and be merry, for to-morrow too there will doubtless be more food." So too the summer will doubtless find flowers for itself; let us set to work again, and, somewhat

10

PLUM TREES

ALFRED EAST

changing Herrick's reason, "gather plum-buds while we may," and cherry-buds, and magnolia-buds, and every sort and shape of catkin which Nature has devised.

March.

AMID the branches of the silv'ry bowers
The Nightingale doth sing: perchance he knows
That Spring hath come, and takes the later snows
For the white petals of the Plum's sweet flowers.

Steel.—From Chamberlain's "Classical Poetry of the Japanese."

March 3rd.—At last! In spite of gloomy forebodings that there would be no trees left to flower, the time of the full blossoming of the plum-trees is at hand, and the gardens are beginning to dress themselves for their long festival. The great trees, pink, white, and occasionally crimson, have begun to light up the dun-brown landscape, of which we are quite weary by this time. Boughs of the crimson Peach are wanted for the decorations for the Girl's Festival—*Jomi no setsu*—which is held to-day. The Plum shares with the Nightingale the honour of being the herald of spring-time ; its buds are the first to yield to the genial influence of sun and shower, and receive from this nation of nature-worshippers abundant and appropriate homage. The artist paints their charms, the poet sings them, calling them "the first of flowers," "the eldest flowers of mother earth."

11

Everything that is white and bright affords him subject for his graceful imagery.

> *THE flowers of the plum-trees*
> *All through the day make snowlight,*
> *Moonlight through the night.*
> *Like the ice-spray which the breeze*
> *Scatters from the stream,*
> *Like the snow-flakes' flight,*
> *Falling petals seem.*

Many a temple is famous for its ancient trees, and takes rich tribute from the people who come to gaze, write poetry, and drink tea under their boughs. Age never seems to make them weary of flower-bearing; year by year their blossoms increase in multitude and beauty, wreathing the old and withering trunks in white and pink robes, as it were for an annual wedding with the youthful year. Like the river, they seem to be endowed with life perpetual. The fashion of the old order may change, the grace of the old national dress may give place to the crude outlines of occidental garments, the country may flourish under its new Constitution, and the people delight in vain things at election time; but let us hope that lovers will still be simple enough to hang their verses on the flower-laden branches, and maids still coy enough to don their prettiest old-time raiment wherein to wander forth to read them, and wonder how they came there. The plum-tree at home, the bearer of Nature's first flower-offering to the family circle, always finds a warm corner in the Japanese heart. *Cettria cantans*, the Nightingale, hides and sings among the flowers, and thus sight and hearing are both gratified.

> *SOME friends change and change,*
> *Years pass quickly by,*
> *Scent of our ancient plum-tree,*
> *Thou dost never die.*

12

Home friends are forgotten ;
Plum-tree blossoms fair,
Petals falling to the breeze,
Leave their fragrance there.

Cettria's fancy too
Finds his cap of flowers,
Seeks his peaceful hiding-place
In the Plum's sweet bowers.

Though the snow-flakes hide
And thy blossoms kill,
He will sing, and I shall find
Fragrant incense still.

Later in March.—Buds, still buds ; and, except for
the evergreens, not a leaf to be seen : but of buds an ample
supply of *Mume, Pyrus,* and Magnolia. At a ball given in
commemoration of the Constitution, with the exception of
a few early Camellias, the decorations consisted entirely of
branches of Pine and boughs of buds, all arranged with the
most consummate taste. One of the hanging baskets was
made of split bamboo, of a truncated 8-shape not unfamiliar
to us in the West ; within the basket an upright glass vessel
full of water in which gold fish disported themselves, and inside
this again a dark green joint of Bamboo holding branches of
Camellia—a thing of joy to be remembered. A few days
later arrive pots with dwarf trees simply smothered with
bloom—among them the lovely white *Pyrus japonica*, almost
a stranger, I think, in England. A few days later still the
Camellia season is full upon us. Of course we know all about
Camellias at home ; I have even seen moderate-sized trees
growing out of doors in England ; but in Japan *C. japonica*

13

GARDENERS GOING TO A FLOWER FAIR WITH DWARF TREES.

grows in the most reckless profusion, and in protected spots
to an extraordinary height ; it is indeed the queen of flowering
shrubs. The wind even is not now without 'a charm, for it
drives the flaming petals like crimson dust-clouds before it,
and piles them in heaps by the wayside ; the air at a village
wedding is not fuller of flung rose-leaves. And where the wind
does not intrude, in secluded villages where the trees grow
more in peace, the streams run seaward laden with a rich red
burden. I think it is the Camellia that first brings home to the
mind the fact that we are living under the influence of a different
climate ; the plum-trees as they are grown here were of course
quite new, but these wild giants covered with flowers impress
one more, and make one wonder whether, after all, the biting

14

A DWARF TREE OF PRUNUS MUME

cold and piercing winds are only the eccentricities of a climate which is in truth more genial than our own.

I have the wish, but not the power, to devote some space to the delightful little flower-trees which are so characteristic a feature of Japanese gardening science. I have, however, been quite unable to get any reliable information as to how the wealth of flowers is produced, beyond the fact, which will probably suggest itself to gardening minds, that the root-growth is retarded by tight straw bandages. The accompanying illustration of a dwarf plum-tree will give some idea of the great beauty of these little trees. The trunks are often thirty, forty, or more years old, but they are covered with a cascade of blossom, the young flowering branches being arranged with a profound art. The science of this arrangement is traditional, and, I believe, unwritten. It is based, of course, on the predominant idea of all the flower arrangements, which is—beauty of line. The trimming of the buds is so arranged that you seem to see every flower on the tree at once—no one standing in another's light.

By this time, the end of the month, which has been a very lion among the months, March seems to have spent his rage and has begun to don his lamb's clothing. The sunny air is full of the accumulated scent of a thousand Daphnes (*D. odora*), pink and creamy white. Every garden is full of the small shrubs, and every shrub is full of flowers. Someone has said somewhere that in Japan the flowers have no scent, the birds no song. There never was a more hasty or a falser generalisation. The Daphne and the Nightingale would alone belie it, and they are not alone. The garden grows daily in colour. I have to note four shrubs all new to me, all brilliant with yellow flowers, and at present all leafless. The *Mitsumata*, or three-pronged-fork paper plant (*Edgeworthia papyrifera*) with its golden balls, from which the Japanese make their tough fibrous paper ; the jasmine-like *Forsythia suspensa*, its boughs simply smothered in bloom ; the pendulous *Corylopsis spicata*, and *Stachyurus præcox*, the " Cowslip tree," as some have christened it ; also the cousin of

16

the Chinese "yellow Plum," *Hamamelis japonica.* Of Camellias, the cry is still, They come: in many varieties of *Camellia japonica* (*Tsubaki*), the chief of which are a small compact pale pink flower, the common brilliant red single one, and a double variety of this with larger petals of an intensely deep crimson. And of *Boké*, the charming *Pyrus*, four varieties: the crimson, so familiar to us at home, which grows here into vast shrubs; the pure white; and two intermediate pinks, the darker a beautiful flower of very regular shape. Lastly, for the houses, yet another *Pyrus*, which with its long stalks seems to be almost a Cherry—*P. spectabilis*—the *Kaido* of the Japanese, which they call the "noble flower," whose praises the poets never weary singing, and whose charms maidens never weary worshipping.

Here is a poem to its honour, *Hana wo netamu*, which the translator has christened

JEALOUS OF A FLOWER.

A THOUSAND drops lie glist'ning,
Fair relics of the night,
Soft rain has gemmed the petals
With pearly tears, and bright
The flashing diamonds display
All sparkling, mirrored by each ray,
Morn's glorious, gladsome light.

'Fore all the noble Kaido
Stands radiant in the glow;
Each blossom gently trembling
Appears to smile, as though
It fain would speak and coyly sigh
For mortal love that passes by
The flower that pines below.

A maid steps from her chamber
  Where winds the garden way,
And wanders, idly dreaming,
  Dreams of a morn in May;
And as she slowly homeward wends,
All graceful from her hand there bends
  A dainty Kaido spray.

The maiden seeks her mirror,
  And musingly her mind
Compares the rival beauties
  Most glorious of their kind—
The loveliest flower 'neath Heaven, and e'en
The sweetest maid those eyes had seen—
  No answer can she find.

So, turning to her lover,
  With wanton lips she cries,
"Which, dearest, think you fairer?"—
  With laughter in his eyes,
"The flower, sweetheart, the garden's pride,
Quite kills the beauty at my side!"
  Her love at once replies.

The maid in sudden passion,
  Scarce waits the words she'd meet,
But rising, lifts the flower
  And casts it at his feet:
From mocking lips the words burst forth,
"'Tis well, you seem to know its worth;
  This night embrace the Kaido, sweet!"

April.

Early April.—The Plums being over, I note the following varieties which have done common service as flower-trees at different times during the past months. I give the Japanese names, as illustrative of the botanical knowledge of the Japanese.[*] *Prunus persica* (peach), deep crimson double flower — *momo*, white variety, *shiromomo ; P. communis—sumomo ; P pseudo-cerasus—sakura*, with enormous pink double flowers the size of a silver dollar ; *P. japonica*, small double pink, the earliest arrival, *niwa mume ; P. mume—mume*—in four or five varieties. The catalogue gives in all eighteen species. I have jotted down a few of them in the accompanying page from my sketch-book. In giving the name *Nankin-mume*—"the Nankin Plum"—to *Chimonanthus fragrans*, noted in February, the Japanese botanist was not far astray. Lindley gives the position of its family in "The Vegetable Kingdom" thus :— *Rosaceæ, Calycanthaceæ, Pomaceæ*—midway, that is to say, between the Roses and the Apples, which, with the Plums, form part of the Rosal alliance. Another instance of this knowledge I may give here, though somewhat before its time. The Japanese name for *Pæonia montan* is *Botan*, and this word enters into the names given to two varieties of *Clematis* — both *Pæonia* and *Clematis* belonging to the Order *Ranunculaceæ*. In the case of *C. tubulosa*

* "The ancient Chinese system of medicine in particular, which was followed until the Restoration, required a knowledge and precise identification by name and form of hundreds of plants ; and a highly developed love of nature, especially a capacity for enjoying beauty in flowers, did the rest. No other people in the world, except the Chinese, has so old and elaborate a vegetable nomenclature as the Japanese. These popular names for so large a number of wild plants are one of the best proofs that an appreciation of nature and keen observation were here very early developed."- *Rein's Japan*.

(*Kusa botan*), this accurate grouping appears to be most remarkable. The Almond (*Amygdalus microphylla*) is also in full bloom here and there.

The gnarled dwarf Pines (*Pinus parviflora*) must not be forgotten ; they are one of the characteristic features of Tokyo, lining all the Palace moats, and many of the roads running by the river. I have three on a rising mound in a corner of my garden, with a stone lantern, to all of which I have done but feeble justice in my sketch.

A CORNER IN MY GARDEN.

middle of April.—The promise of the early spring has been fulfilled ; buds no more delight us, but fuller flowers in plenty have full sway. After a fortnight of cold winds and still colder rain the sun has won his wooing, and the glad earth puts on her bridal veil of cherry-blossoms. And underneath that wondrous veil, the like of which—this I write deliberately—no western mortal has ever seen before, she has completely changed her vesture. The dun-brown dead grass has passed like a dull and ugly dream away; a new spirit has come, like the bridegroom in the night, over the landscape, and one wonders if it really be the same, or whether the whole scene has changed. Curious effect of the delight a Japanese scene evokes; it is not only a joy for ever, leaving in its train a hundred tender memories, but it blots out, as the daylight blots out dreams, the very recollections of wind and driving rain; those too turbulent elements cease to have had existence. What wonder that man's ideas in this strange forgetful land have been fashioned by Nature's teaching! As for the past—was it dull and drearily uncomfortable, *shikata ga nai*; it is dead and must bury its dead : as for the present—it is beautiful, let us enjoy it.

Tokyo is one vast garden. In the painting of her picture, Nature has laid on her colours thickly, and with no sparing hand; the grass below and the trees above are all bursting—no, that does not adequately express what is now taking place— rushing, racing into leaf. Here, great patches of white cherry-blossom; there, great patches of pink cherry-blossom; and, in between, a flaming crimson Peach. Camellias, first in the flower-race, still linger to see the sight they preluded; ever and anon Magnolia trees, leafless as yet, with blossoms standing up like great candles, which seem to make the daylight linger and live long into the night; and lastly—sight of all sights, the strangest and most new—the budding Maples floating across the landscape like crimson clouds in the sunrise of a dream.

"There is snow lying thick on the Yoshino mountain." "A tinted waterfall comes headlong from the sky." "A white cloud hides the mountains." But there is neither snow, nor water, nor thick vapour; they do but live in the poet's imagination as he looks at the blossoming cherry-trees. And on this mid-Sunday of the fragrant month, every woman in this land of fair and gentle women dons her finest raiment and her gayest dresses, and goes forth as it were to a wedding festival. In the temples where the trees are, the gods who gave them, or

21

the ancestors who planted them, are roused by many a tinkling bell and gong, that they may share in spirit-land this most divine of earthly joys. Merry, laughing, pattering, toddling, nonsense-talking people, may you keep this happy festival for many and many a year to come, through all those "ages eternal" of which your Constitution so boastfully speaks (in English; though in Chinese it speaks more soberly of the ages eternal of the past); and when the order has changed completely, and all that is yours of loveliness and grace has passed utterly away, may this day, with its many consorts of yesterdays and to-morrows, keep green the festal memory of old Japan.

TO A MAIDEN OF OLD JAPAN.

[In very irregular metre.

*M*AIDEN *childlike, in thy hands*
　　Old Japan must live,
When the customs of strange lands
　Curious blessings give.
When thou'rt learned in foreign ways
　Think thine own the prettier;
Fragrance in the memory stays,
Fragrance of the vanished Mays;
Mark, then, all these cherry-days
　With the reddest letter.

Dainty breath and dainty finger
　Keep the dust and moth away
From thine old-world vesture.
　Let thy dress of soft cloud-grey,

22

With its silken texture,
  Be the fairyland and gay
  Where old Japan may have full sway,
And memories fondly linger.

New Japan! No softer hours
Than those we spent among the flowers
The Gods can ever send her.
Then once a year forget the West,
Take from thy closet what is best ;
Go forth in flowered garments dressed,
Thy form by all their folds caressed ;
  Brave in them the blossom-showers,
Then lay them by in lavender.

The gentlest breeze, away the flowers fly, making the air glitter in the sunlight, and the girls cower and bend their heads as it were before a gusty snowstorm ; and then they brush the petals from each other's shoulders, calling them snowflakes, and the air is full of the ripplingest laughter that ever was laughed in the land of make-believe. And then, how strange! what food for more light laughter! the more the petals fall, the whiter look the trees, and still more full of flower. Those that have fallen have but made room for those that are left behind to be seen, to feel the breeze, and fall. Again the simile of the snow : "The drift looks whiter as the wind blows away its outer coating in clouds."

A petal of cherry-blossom sayings picked in passing :—"One day Kinto Fujiwara, Great Adviser of State, disputed with the Minister of Uji which was the fairest of spring and autumn flowers. Said the Minister : 'The Cherry is surely best among the flowers of spring, the Chrysanthemum among those of

23

autumn.' Then Kinto said, 'How can the cherry-blossom be the best? You have forgotten the Plum.' Their dispute came at length to be confined to the superiority of the Cherry and the Plum, and of other flowers little notice was taken. At length Kinto, not wishing to offend the Minister, did not argue so vehemently as before, but said, 'Well, have it so; the Cherry may be the prettier of the two; but when once you have seen the red plum-blossom in the snow at the dawn of a spring morning, you will no longer forget its beauty.' This truly was a gentle saying."

If the blooming of the flowers touches to the quick the susceptibilities of the Japanese for beautiful things, their fading appeals to the sentimental side of their nature. "The cherry-blossoms," says one, "are ineffably lovely; but my joy at gazing at them is marred by the knowledge that they must so soon pass away." And the poet Korunushi sang

" *NO man so callous but he heaves a sigh*
When o'er his head the withered Cherry flowers
Come fluttering down. Who knows? the Spring's soft showers
May be but tears shed by the sorrowing sky."

From Chamberlain's "Classical Poetry of the Japanese."

Just now the round of flower-visits is in full swing. Some Westerns have been known to wonder at the time spent upon them, and to regard them as a purely Japanese craze, worthy only of an indolent, time-oblivious people. Yet I have seen—or have I dreamed it?—advertisements on the Underground Railway at home, directing public attention to the fact that the Chestnut trees at Bushey Park were in full bloom; and, more curious still, I seem to have some dim recollection of people—British people—going to see them. If these things be so, then truly is there nothing new under the sun, not even the Japanese sun. These delightful flower-days even have their counterpart in

24

IN THE TIME OF CHERRY BLOSSOM

ALFRED EAST.

the old world; the only difference is, that in Japan they are more elaborated, more carefully thought out, with a view to the extraction of the greatest amount of pleasure *sans* alloy.

A brooding philosopher says of these flower excursions, half regretfully, "If there were no charming cherry-blossoms in the world, the mind would be left undisturbed in spring-time."

Yononaka ni taete sakura no nakari seba, haru no kokoro ya nodoke karashi.

The curious twist of the Japanese mind reads between the characters, and says he is no philosopher at all who speaks, but a disappointed lover railing at woman and her disturbing presence in the world. Such a lover as he who once thus cried to his cruel fair : " The cherry-blossoms have come and passed away without my seeing them, the while I am trying to win your love." Your true philosopher, however, thus points his moral to those less brilliant ones who have to wait their turn until the lustre of others' fame has faded : " The cherry-trees in the far-away mountain villages should keep back their blooms until the flowers in the town have faded, for then the people will go out to see them too."

Miru hito no naki yamazato no sakura hana hoka no chirinan nochi za sakamashi.

April 20th.—To-day the sunlight and the grass have begun to work their never-failing charm, and my mind is full of memories of green lawns at home. Even this strange, straggling, rootless product of the Hermit Kingdom is potent with the spell; and the brilliant *Kerria japonica*, now in full orange bloom, helps to fill in the mind-picture with cottages in the balmy South of England, over whose trellised portals the charming flower had spread. It was a domesticated homely flower as I knew it, long

E

ago, before I could even have spelled "Japan;" and has entered into the flower memories of earliest days. I grasp its long tendrils, shaking them as the hand in the greeting of old friends.

With an old a new friend makes his appearance, one destined hereafter to form part of every thought of Japan. Ruminations in the wander round my garden are suddenly cut short by the appearance of a thick spike, some eight or nine inches high, which seems to have pushed itself through the turf in a single night. After a while I notice many more, and the ground seems full of spikes of lesser magnitude in process of mushroom-like development. Something quite new! my slender stock of botanical knowledge suggests Aloes—at the mention of which, in the Japanese tongue, my gardener laughs, but vouchsafes no intelligible explanation, wondering at ignorance so profound. On the morrow, when I notice at least two inches growth in the shoots of yesterday, and many new spikelets, I learn at length that it is *Také*, anglicé, Bamboo; and that these are the young shoots of the plant which, as a staple, plays a part in Japanese economics as important almost as rice. It would be an easier task to say what is not made of Bamboo, than to convey any idea of the thousand-and-one ways in which shoots, stalks and stems great and small, joints, twigs, scales and leaves, are used by this astonishing people, who not only know how to make the best of things which nature provides, but do better still. My diary is, however, botanical and not economic, and I note, therefore, *imprimis*, the simply astounding rapidity of growth, which transforms a shoot into a tall plant in a few days. The tall stems seem to attain to their full height almost as quickly as the smaller ones. I fancy, the accurate way of describing the different growths would be to talk of the shoots from the old, or the young, "leaders" under ground; one is rather disposed, nevertheless, to apply the words old and young to the shoots themselves. The stems do not increase in diameter after they emerge from the ground; they are covered with dark-spotted scales folded one within the other, a scale for every joint. They grow like whips,

26

YOUNG BAMBOO SHOOTS

F T P

without a sign of branch or leaf until three-fourths of their full height; then, the cold nights having quite gone, the stem needing no more protection, and the embryo branches being fully set, the scales drop off one by one, revealing the delicate green stalk, and alternately at every joint a tiny branch. For another fortnight the growth of parent stalk and branches goes on rapidly; but there is no sign of leaf until the full growth is finished, when they look like so many waving and branched fishing-rods. Then suddenly, after a few days of sun and shower, the stalks are hidden in a wealth of foliage of the tenderest green, which waves to the slightest breeze, making a delicious picture with softest lights and shadows, an ideal fairy grove.

April 26th.—After a feast of double cherry-blossoms in the Imperial gardens, I find in my own garden, with an abundance of flowers, *Aquilegia atropurpurea*, purple and white varieties of the ever-welcome Columbine; *Epimedium macranthum*, a berberid, but not unlike the Columbine in form; *Magnolia obovata v. purpurea*, a very magnificent bush with great dark purple flowers; the white and the lilac *Fuji*—*Wistaria chinensis*, of which more anon; and My Lady's Slipper, *Cypripedium japonicum*.

April 28th.—How can I describe a Japanese garden? Words fail me when I attempt to portray its delicate delights. Yesterday a quiet ramble through Count Okuma's garden at Waseda, on the outskirts of Tokyo, revealed, as I had never seen them before, the thousand little charms which go to make up the sum of any single day of pleasure in Japan. As you step from the verandah, the calculated convenience of the stepping-stones first, and then the gold-fish in a huge Chinese tank of clearest water, fill in a pleasant moment. The lawn has never yet known the whirr of the patent mower, but is kept soft and smooth and trim by an army of men, who minister to its daily sleekness with small reaping-hooks and scissors. Walking delicately, you come

29

to the dry watercourse, its stepping-stones laid with marvellous
care. With an eye to effect ?—no ; apparently there is no effect
at all, everything seems so natural and in its place ; the only care
has been to efface all trace of effort. Every stone has been
selected for shape and colour and fitness, the large ones at a
great price. Of course, the soles of European boots leave horrid
marks behind ; one longs to stop and rub them away. The
minute labour of crossing the stream accomplished, a clump of
tree Pæonies—*Botan* (*Pæonia moutan*)—invites you to dwell
awhile before you climb the tiny Fujiyama beyond. Then a path
set in a graceful line, and hedged with many-tinted Maples (*Acer.
palmatum*, of which, if my memory serves me right, there are
seventy-eight varieties in the garden), green and white, orange,
crimson, brown, leads you up the miniature precipitousness of
the mountain, modelled after that peerless one, the worship of all
Japan. On its summit a shrine and a resting-place, from which
you may look back on the beautiful simplicity of a Japanese
house, a harmony in brown wood and cream-coloured mats, with
here and there a flash of light as the sun strikes on some half-
hidden gold screen or sliding door.

Next, a shady glade of trees and flowers, leading to a rest-
house, wherein is served minute refreshment after so much
minute labour—tiny cups of scented tea and oranges. What a
gem of a house ! what a picture ! with its adornments of
fantastic natural wood, and delicately-toned screens. Everything in most delightful miniature: its curiously twisted bamboo
fences, its stone steps, its bowl of gold-fish ; the gold-fish themselves ; even the time of resting. The only thing not in little is
the infinite labour that welded so many minutiæ into so charming a whole. Still more glades, more bridges bosomed in Maples,
more trees, more flowers ; and then there comes one of the most
charming features of a Japanese garden—the sub-tropical greenhouse. It is the simplest thing possible in construction : four or
five tiers of benches some fifty feet long, with ample space
between, screened from the north by bamboo fences, and for

shelter from sun and rain movable mats stretched out on a slender framework. On the benches, flowering plants and trees of curious growth, each in a pot of great beauty and value. The difficult question, which is to be admired most, the plant or its pot, is settled for once in the Japanese manner. If the plant is to be specially admired, the pot ought not to detain you ; if the flower is modest, the beauty of the pot will complete the picture. Above all things, the characteristic of an English greenhouse is wanting : there is no crowding, no confused hodge-podge of colour ; each occupant has a particular space devoted to it ; he claims your notice, and he gets it, though it be for a moment, all to himself : his beautiful comrades will get theirs in turn. Then back over bridges and under trees, past the gold-fish bowls once more, over the spotless mats, and past the priceless screens, and so *sayonara* to the statesman and his flowers. The blooming of the *Botan* at Waseda has been like the blooming of the Roses at Hughenden—a consolation in the hour of bitter defeat.

May.

May 5th.—The festival of boys—"*Tan-go*"—and of the early Iris. The Iris is supposed to be a specific against all disorders, a charm against all evil spirits. The bath-houses provide Iris-water for the bathers, and the superstitious hang bunches of leaves and flowers from the eaves of their houses. Probably the same superstition led to the common custom of planting beds of Iris along the ridges of the thatched cottages in the country. In days gone by, boys wore wreaths of Iris leaves, and made ropes of them to dance with, and beat the ground to frighten away the demons from their festival.

May 14th.—Within the last week three famous gardens have been at their best, and crowded for many successive days with the merry laughing Japanese crowd which is so delightful when one is in a good humour; and it must be a very bad temper indeed that cannot put off its churlishness in the presence of so much spontaneous, unaffected gaiety.

At Okubo, a flower village on the skirts of the capital, where dwell market-gardeners innumerable, are fields and fields "brocaded" with dark crimson Azalea (*Rhododendron indicum v. obtusum*); every tree is a thick mass of blossom; many of them are old, and apparently long past flower-bearing, but they are more prolific even than their youthful offspring. In the gardens choice specimens are grown of every shade of yellow, salmon, and tawny rose.

At Shoko-yen a tiny garden nestles in the sheltering hills, its owner devoting himself to the culture of the tree-pæony; a hundred varieties, wonderful in size and colour, are in full bloom.

32

A WISTARIA ARBOUR.

F

Lastly, there are the Temple grounds at Kameido, where the white and the lilac Wistaria grow in astounding profusion. The racemes of lilac-flowers are hardly ever less than three, and often five feet long, and they hang down from the trellis work in a dense mass of colour. "The reflection in the lake makes it look," says the poet, "as if from its bottom there also grew a harvest of flowers." Under the many trellises the world sits and sips its tiny cups of tea, gazes at the fish in the ponds, wonders at the beauty of the flowery tassels overhead, and laughs at the unsuccessful attempts of others to get gracefully over the semi-circular bridge. Then the frivolous maiden buys herself a new hair-pin with pendant Wistaria, in memory of the pleasant day. The white *Wistaria* (*W. brachybotrys*) has larger flowers but much shorter racemes than the lilac, and is not of such free habit. Arbours are, however, often covered with it. Used as a dwarf tree for the house it is exceedingly handsome.

The Plum and the Nightingale are the harbingers of spring, the Wistaria and the Cuckoo of early summer. Hitomaro, an old Japanese poet, sings of them thus :—

IN blossom the Wistaria trees to-day
Break forth, that sweep the wavelets of my lake :
When will the mountain Cuckoo come and make
The garden vocal with his first sweet lay ?

From Chamberlain's "Classical Poetry of the Japanese."

Yet another poet likens the fragrance of the drooping flowers to true friendship : "What," says he, "though I be outside the ring-fence and cannot sit beneath thy shade, thou sendest, gentle Wistaria, thy fragrance across it to me, treating me like a friend."

In the garden at home Lilies of all sorts are sprouting vigorously, and the early Iris shows its bloom ; also the rich

34

ARRANGEMENT OF WISTARIA AND IRIS.

orange *Lonicera sempervirens v. minor*, which we call Japanese
Honeysuckle ; *Spiræa cantoniensis*, a charming, compact white
flower quite unknown to me ; *Polygonatum canaliculatum* and
its minute companion, *P. humile ; Bletia hyacinthina*, a hardy
orchid growing in clumps with many flower spikes, also in two
varieties, bluish crimson and white ; and *Asarum caulescens*, the

35

tiny flower whose leaves furnished the great Tokugawa Shoguns with that famous crest which glitters on a thousand shrines and temples. Best of all, the great bare poles of the *Paulownia Imperialis* are just bursting into leaf and flower, the imperial *Kiri*, which furnishes the second of the Emperor's crests, sharing the honours with the *Kiku*. Seen to perfection in full flower, and when it has not been bullied by the wind, it is an exceedingly handsome tree; clumps of them on the warm hill-sides, with their great spikes of flowers, tinge the landscape with a violet hue. The leaves of the young trees are something prodigious: a sapling five or six feet high will bear leaves often more than eighteen inches across; this is probably due to the extraordinary vitality of the tree. Lately I used some old pieces of a tree, uprooted for three years, in making an arbour; they had not been in the ground a month before they began to sprout, and were soon covered with small leaves.

Mid-May.—I have reserved my notes on the night flower-fairs until to-day, when the Roses are in bloom, for the Rose fair is the prettiest of the year. First, however, the quaint story of the introduction of Roses into this country.

In the early days when the West was as interesting to Japan as Japan is even now to us, the Japanese showed their aptitude for entering the communion of the civilised States by starting crazes for European things. Three famous crazes still dwell in the memory of men—Pigs, Rabbits, Roses; and stories are plentiful of the almost fabulous prices paid for them. I am afraid to trust myself to figures, but two to three hundred dollars for Rose, or Rabbit, or Pig, is, I believe, well within the mark. The Rose seems quickly to have found its way to the Japanese heart, as well it might, for all hearts at all times have paid it tribute. Prices went to an extraordinary figure. There lived then a man in the city who saw a road to fortune strewn and scented with rose-leaves; the savings of a lifetime were invested, the garden planted with trees paid for literally with the weight of their flowers in gold. The sun shone, they were digged and trenched, and the Roses bloomed. But the favouring gales brought other ships freighted with the fragrant flower ; then came the collapse ; the roseate bubble burst ; the price so fell in a single day that one could scarcely write *sen* in the place of *yen*. The rose-trees went on blooming unconscious of the evil they had wrought, but there was none to tend them, or gather their flowers. The ill-fated gardener had vanished ; no one knew where to set up a floral tribute to his memory, and none but the birds where to scatter rose-leaves at his burial.

Most curious are these flower-fairs, which go on all the year through. Each district has one every month, when its inhabitants take a holiday ; many of the days being fixed so that the evening fair may come as a pleasant termination to the festival at the Temple hard by.

The gardeners have thus a regular succession of markets to attend, but I am bound to say I have never yet understood how, after a short time, they have anything left to sell, for they bring large trees and shrubs for sale besides their innumerable plants. Of course it is something like the stage army, the larger trees moving on from market to market until they are disposed of ; but the impression left on the mind is, that the majority of the gardeners are being sold out of house and home. And yet the supply goes on all the year round. The most amusing part of the business is, first, the exuberantly extravagant price asked for the flowers ; secondly, the ridiculously diminutive price actually taken. It has indeed passed into a proverb, for the custom is not reserved for foreigners, but is applied to Japanese as well, that any tradesman who asks high prices "charges like a florist at a festival." But one is compelled again to wonder, after a few fine sturdy plants have been bought for a few *sen*, how the peripatetic florist can live, for it must be with difficulty that he can sell a dollar's worth in an evening.

But it is always the same in Japan—the difficulty of finding out the true value of anything; "what it will fetch" seems to be the only rule. Two commercial ideas seem alone to prevail. If anything has cost money, it must rise gradually in price so long as it is unsold, in order to bring in interest on the original outlay; the older the stock, the higher the price. Secondly, if only that worthless commodity, labour, has been expended in production, the price really is unimportant so long as you get something. The only result is that one hour's work may bring in less or more than another, and, what can that matter?

With this perpetual round of fairs the Japanese house need never be without its branch or pot of flowers. To the foreigner they give the first insight into the astonishing skill in transplanting trees, which the Japanese gardener possesses. Nothing is too big to move; they know the times and the seasons for doing it, and the way moreover to do it successfully; transplanting a garden is indeed as much in a day's work as planting it. The law and the custom have grown up side by side; under the old law, which remains unchanged, trees and shrubs are movables, and belong to the tenant who planted them; they are part of the slender stock of household gods which he carries with him in his frequent migrations. At certain times of the year whole copses come upon you, climbing up the hills and wandering through the streets.

June.

The Early Days of June.—Grass cutting at the rate of about four square yards, or less, *per* man *per* day.

Of flowers : *Spiræa japonica*, two varieties, light and dark pink, in masses ; *S. purpurea*, nearly allied to Lindley's favourite flower ; and *Astilbe japonica*, the old white *Spiræa*, so abundant in greenhouses at home. Twining over the arbours the rich purple *Clematis florida*, the same size, but a fuller colour even than *C. Jackmanni* ; and its delicate white companion, *C. florida v. alba*. *Lilium longiflorum*, the handsome white gun-shaped Lily (japonicè *Teppo yuri*) ; *L. Thunburgianum*, in several varieties, one especially handsome, with large crimson flowers dashed with orange ; also the beautiful yellow " day-lilies," *Hemerocallis flava*, the brilliant orange *H. fulva*, and a small-flowered variety of the same family, probably *H. minor*.

Seeds have produced the handsome thistle-like orange *Cartha-mus tinctorius*, and the delicate little *Acorus gramineus*. And among " weeds," *Lappa major*, *Chenopodium album*, noticeable for the lustrous red of the new leaves of the young plants ; *Potentilla fragarioides v. stolonifera ;* and in great abundance everywhere the curious white-flowered *Houttuynia cordata*.

A country walk in the mountains round Miyanoshita, full-bosomed with dark green trees suggesting depths unfathomable, and pine-trees standing up " islanded in the immeasurable air ," then a break, and a waterfall tumbling down the green hill-side.

The banks of purple Iris on the roofs of the cottages which glisten in the sun are in full bloom ; and by the wayside are great clumps of *Iris japonica*, its pale grey flowers lit with bright spots of purple and orange ; dwarf *Wistaria* peeping out in odd places ; some small varieties of *Spiræa ; Diervilla versicolor* and *D. grandiflora*, of the tribe and lineage of the Honeysuckles ;

39

Deutzia scabra, the old familiar friend ; and growing by the active sulphur-beds of Ojinoko, a charming little family of hardy Orchids, *Calanthes* and *Cephalantheras*. Lazily drifting on Hakone lake in the shadow of Fijiyama, we seem to pass into the eternal depths of the ever-present mountain. *Per contra:* heat, rain, fleas and mosquitoes.

June 20th.—Before the wind, remorseless, blows out the lantern-candles of my memory, and scatters, like flower-seeds, the impressions of a lovely little trip up the Sumida river, I must try to get on paper the recollections of it. How the wind howls! all days are not calm in Japan, though they glitter with gold sunbeams. No poet ever sang a welcome to the wild wind of Dai Nippon : I suppose, because poetical necessity demands that the subject apostrophised should at least be knowable by some name, and he would be a hardy poet who should attempt to localize the quarter of the globe whence come these gusty billows of impetuous air. Yet another name for these islands of the East—"The Land of the Rising Wind."

On this day, then, full of golden sunlight, the wind wafted us (though that is but a foolish word wherewith to express the force which it expended on the sail, driving the boat before it) up the Sumidagawa towards Hori Kiri—the "beautiful ditch"— in which the lovelier and loveliest Iris have their dwelling. Half an acre : in which room has been found for a tea-house on an island, a rest-house on a knoll, a long winding path lined with bright yellow day-lilies to a rest-house on a hill, meandering paths to a dozen smaller rest-houses, and yet room enough for the watery bed of a thousand purple Iris here, and for another of a thousand white Iris there, and in the remaining space thousands more of every tint from purest white, through the purples, down to pale crimson. Large flowers of many hues

40

HORI KIRI.

have been seen in Tokyo for some time past, but the size of these in the beautiful ditch is extraordinary. "How shall I tell how many Mays the Iris flowers have lived through? Every year they look younger," as the setting sun, glinting their petals, lights the coloured fires of the fairies' home, through which we can see them waking from their day-dreams, one by one, to join in the sunset dance.· As if by magic the turbulent wind has fallen asleep, and we walk back to the boat through the paddy fields, glowing as though the earth had opened her mouth to show the great furnaces below. The little maidens of the riverside tea-house busy themselves with our embarcation, which, being effected, becomes the sum total of a hundred sweet little nothings done with a grace divine. Then, with lighted lanterns casting pale-tinted reflections on the water, with "*sayonara, sayonara,*" sweetest of all sweet Japanese words, softly and still more softly dying away in the darkening distance, to the tinkling music of the Samisen we drift away—to dine.

June 22nd.—Hydrangeas: *H. hortensis v. Azisai,* the ever variable, now pure white, now electric blue, and changing to red as its clusters fade ; *H. hortensis v. japonica,* whitish-green centre with red external florets ; *H. virens,* blue centre, with white and pink florets.

Lilies : *L. concolor v. pulchellum (Hime yuri,* the "young lady lily " of Japan), a small, rich orange flower ; *L. medeoloides,* the "wheel" Lily, so called from the whorl of leaves thrown out at the first joint.

Generally : *Punica granatum,* single, double, and variegated Pomegranate; St. John's wort in several varieties; *Funkia lanci-folia,* with a spike of dark purple lily flowers, its pale taller brother, *F. Sieboldiana,* both brought down from the hills last year, and a gardener's variety of *F. ovata* with handsome variegated leaves. Also, from the hill gardens, a few plants of the crimson *Lychnis grandiflora,* and from the hill-side hedges its Indian-red brother, *L. Miqueliana.*

42

The golden clusters of tiny flowers of *Diospyrus kaki* give promise of a rich harvest of luscious fruit in the autumn—the *Kaki* loved of the Japanese.

ARRANGEMENT OF LEAVES, BUDS, AND FLOWER OF IRIS.

July.

July 7th.—Plots of grass everywhere pink with the tiny spiral Orchid, *Spiranthes australis.*

On this day occurred in times past one of the most graceful of the fast disappearing customs of old Japan. The Goddess, whose dwelling was in the Star "Vega"—"The Weaver"—whose care was for all women who lived by the needle, was wont, on this day, to meet her lover—"The Herd Boy"—in the Milky Way. Then all those women worshipped them, invoking the blessing of these happy lovers on their embroideries and on all kinds of needlework. At the close of the day they made them little rafts of wood, fixing in each a votive spray of flowers, and sent them drifting down the river.

July 10th.—Summer heats have begun to make work impossible, and to drive people to the mountains. Tokyo moves itself somewhat arduously, bag and baggage, chairs and bedsteads, to Hakone or Nikko. For me Nikko, dampest yet pleasantest of places. The train leaves before seven, the station is five miles off, and that means rising at half-past four. At four the vigorous patter of rain-drops wakes one, only to encourage further sleep and forgetfulness; but visions of a day in Tokyo, with cook and luggage gone before and the house packed up,

44

suggest that even a long damp journey is not the greater of the two evils. Energy meets with its reward. Before the railway journey was finished the clouds had packed themselves away, and the sun shone out gloriously, making the five hours in a *jinrikisha* tolerable; for in the sunshine not even a foot of mud, and consequent jerkings innumerable, can destroy the indescribable grandeur and charm of that avenue of Cryptomerias, which looks from the Nikko hills like a great dragon stretching itself lazily through the landscape. The new railway has now brought peace to the giant trees, and has taken the Nikko traffic off the road. But the sacred avenue has been desecrated; the iron road cuts through it in two places.

On the morrow, a ramble over the wonderful hills, whereon are great bushes of crimson Azalea (*Rhododendron indicum v. macranthum*) still in flower, peeping out from between the dark Pines. "Are they flowers, indeed?" says the poet; "I thought they were the woodmen's fires upon the hillside:" "I thought that the mountain goddess *Sao-hime* had passed by, and I had caught a glimpse of the scarlet skirts of her raiment:" "I thought the red-leaved autumn had come before its time."

In the village gardens is *Lychnis grandiflora* transplanted from the hills. Under cultivation the flowers often exceed two inches in diameter, and their natural crimson intensifies to blood-red. But it is still early for the hills; and though there is an abundant promise of beauty in the hedges, at present there is little else but *Actinidia polygama*, with flowers to be mistaken for orange-blossom; and the double *Deutzia* (*D. scabra*), which sheds its delicate petals like snow-flakes in the way. Again the poet sings of moonlight, snow, and waves, as he looks on this charming flower; and yet again, as he looks at the lanes, lined with the white bushes, he likens them to the long strips of cotton stretched out to bleach, which are so common a feature in the summer landscape.

45

WHO hung that cloth so fine
*   To bleach in the white sunshine ?*
Who owns that cloth so rare,
That only a maid should wear ?
Nay, but we are dreaming :
It is the moonlight streaming
Through the hedges and the trees ;
The summer is not glowing,
'Tis winter that is blowing
Those snow-clouds on the breeze.

July 20th.—Notes of a country walk. There will be now
nothing but a diary of country walks, not through meadows, but
over great plains of flowers changing week by week, now white
with Hydrangea, now crimson with Lychnis, now purple with Iris.
To-day the plains are white as with the bridal-veil of summer :
the wreath is of Veronicas, small feathery Spiræas, "Meadow-
sweet" making the air redolent with scent of home, and clumps
of Hydrangea (*H. paniculata*), pure white in the hills—all this
whiteness dashed with the delicate colours of *Spiræa purpurea*
and the graceful lilac spikes of *Funkia ovata.*

In the hedges one Orchid (*Epipactis gigantea*), with a spike of
small orange flowers, varying between two and four feet in
height ; the dark purple star of *Vincetoxicum Nikoense,* and here
and there clumps of dull crimson and white Canterbury Bells
(*Campanula punctata*), together with quantities of bright blue
Commelina communis, and occasionally the scarcer white variety.
The blue Spider-wort is much used in dyeing ; and as it gives
off its colour easily, the fickle mind and transitory love are
compared to it.

46

And then—how can I describe the effect of the first sight of that king among the flowers, the wonderful *Lilium auratum*, wild in its mountain home! As you walk, suddenly it arrests you, as it were a great white eye staring at you through the bushes. Another week there will be thousands of them, but to-day, it is this one, this first one, which makes you pause and wonder at Nature's marvellous handiwork, the delicate stem bending but never breaking beneath its weight of flower, and, hanging over it, a fragrant haze of scent.

July 21st. — Something quite new, and perfect in its beauty—the creeping Hydrangea (*H. petiolaris*), winding round the tall pine-stems and enveloping them with a myriad clusters of white flowers. Also another new flower, *Lilium cordifolium*, growing in damp places, with curious elongated, primitive-looking green flowers, with dark brown centres; the old bulbs, bearing a dozen flowers or more, are exceedingly handsome.

A bunch of flowers from the upper regions:—*Aquilegia glandulosa*, the dark red Columbine; *Scutellaria indica v. japonica*, with purple flowers, growing in clumps by the mountain paths; *Thalictrum aquilegifolium*, and *T. tuberiferum*, very handsome and feathery, with four other species, noted in the catalogue; *Philadelphus coronarius*, with clusters of Syringa-like flowers; the pink *Polygonum bistorta*, and *Wickstræmia canescens v. ganpi*, with its small spikes of white flower.

August.

August 1st.—To Chiuzenji, of the charming lake and pilgrim mountain Nantaizan.   Trees covered with festoons of pale green *Lycopodium Sieboldi*, making charming patches of light among the dark trees.   This curious plant, at a certain altitude, attaches itself to the trunks of the numerous dead trees in the forests, and covers them to their topmost branches.   I have never seen it on a living tree.

The Camellia-tree (*Stuartia pseudo-camellia*), covered with white flowers, is also a noticeable feature on the hill-sides.   Also I note *Adenophora trachelioides* and *A. verticillata*, sisters of the family of Campanulas, purple, lilac, and white, the latter variety often throwing up spikes three feet high ; *Hydrangea hirta*, with its curious colour mixture of vivid blue flowers and bright yellow-green leaves ; *H. involucrata*, with handsome lilac clusters, and of course gleams of their shrubby white brother creeping in all sorts of out-of-the-way corners ; *Tricyrtis flava*, quaintest of the Lily tribe, in waxen abundance, with here and there the white red-spotted *T. macropoda ; Pertya scandens*, twin to the Groundsells ; tall purple spikes of *Phyteuma japonicum*, and crimson *Pedicularis resupinata ;* and above all the regular-petaled orange vermilion cinquefoil of *Lychnis Miqueliana*, which, unlike its crimson relative, is as handsome wild as it is under cultivation.

By the lake, bushes just coming into flower of *Tripetaleia paniculata* and *T. bracteata ; Hydrangea cordifolia ;* also beds of *Aconitum Fischeri*.

A few days later.—Across the beautiful Yumoto plain, which, for all its square miles, is one vast Iris bed.   To-day in the sunshine it is literally one huge purple sheet, patched here and

48

FROM THE NIKKO HILLS.

ALFRED EAST

there with *Lilium tigrinum* and *L. medeoloides*. In the long grass dwell *Parnassia palustris*, *Geranium Sieboldi*, *Epilobium angustifolium*, and the deep orange Groundsell *Senecio flammeus* ; also a spike here and there of the large bright yellow *S. japoni-cus* ; abundance of delicate white Spiræa and the white " Man-orchis " ; by the lake the yellow Spiræa-like flower *Patrinia scabiosæfolia*. In the damp woods, hidden in the under-growth, the delicate tinted cups of *Pyrola rotundifolia*, and the tender-hued pine-root parasites, *Monotropa hypopitys v. hirsuta*, like the pale yellow ghost of a hyacinth, and *M. uniflora*.

Coming back to Nikko, I find the handsome Thistle, *Cnicus purpuratus*, in full flower in the pass, the flowers full two inches in diameter.

August 10th.—There comes over me the refrain of the song, " Oh ! my garden full of Lilies ;" the garden is the broad swell of the mountain side, and the Lilies, both in number and size, exceed those of the gentle poet's dream. The spikes of *L. auratum* with eight and ten flowers, one for each year of life, so the Japanese say, surrounded on all sides by their abundant off-spring, are a sight for weary eyes. The larger heads are rarely found now in the woods : collectors have been too greedy in their search ; but in the cottage gardens fifteen to twenty flowers is no uncommon sight. Yesterday the coolies brought in a spike over nine feet high ; the house for a week was heavy with the scent of forty-two flowers, each of them about ten inches in diameter ; a thing of joy to be remembered ; it cost twenty-five *sen*. Stories are rife at this time of year of even bigger heads of flower, but I think the palm must rest with mine. Investigation showed that most of those that were talked about, with sixty and even eighty flowers, were flat-stemmed ; three or four stems had been thrown up, but the plant had not sufficient strength to separate them. Mine was single and true-stemmed. In the gardens too there are beds of *L. speciosum*, the lovely pink flower we know so well in the West, with two varieties, one a

11

very pale pink, the other pure white with a green vein in the petals, very handsome ; spikes also of *L. Tigrinum (Oni-yuri)* with fifteen to twenty flowers, and its blood-relation, *L. Maximowiczii (Ko-oni-yuri)*, six or seven feet high, with twenty to thirty blossoms. The flowers are hardly to be distinguished, but the stem of the latter is covered with a white wool, and bulbils are thrown out from the leaf bases.

Great trees of *Clerodendron trichotomum* give a new and charming tint to the landscape. Among the lesser flowers, the large white *Funkia* has given place to the smaller purple one, *L. lancifolia*. *Pardanthus chinensis* is in flower, and the hedges are full of the stinking *Pæderia fœtida*, *Impatiens noli tangere*, *I. textori*, and the delicate electric-blue *Clematis tubulosa*. On the mountains the magnificent purple Campanula, *Platycodon grandiflorum*, with pale lilac and double varieties in the gardens. I can never forget the effect of thousands of these beautiful flowers carpeting the hill-side of Bandai-san ; and the terrible transition from the field of radiant blue to the vast expanse of desolation, as we came suddenly to the terrible track of last year's explosion.

To be noted also, that delightful cousin to the Anemones, *Anemonopsis macrophylla*, the pink spike of *Lythrum virgatum*, and (hiding its delicate orange flowers in the undergrowth of the hedges) the Quince, *Pyrus japonica v. pygmæa* ; also in great abundance, *Cimicifuga simplex*, the "Candle-plant," its beautiful spikes of white flower, often a foot long, standing erect above the more humble flowers of the plain : and the mighty creeper, the red-flowered *Pueraria Thunbergiana* covering up every waste spot by the roadside.

From Chiuzenji comes a bunch of quiet, interesting flowers : from the hedges the Blackberry, *Rubus idæus v. strigosus; Pertya scandens, Macroclinidium robustum,* and *Saussurea Tanakæ,* their habits different, but their groundsell flowers alike, those of the last two, in fact, being almost identical ; the two Salvias, *Chelonopsis moschata, Salvia nipponica ;* the pale yellow *Lysimachia davurica ;* the dark-tinted bell of *Glossocomia lanceolata* which creeps over the mountain hedges ; *Aster trinervis*, a Michaelmas Daisy ; *Halenia sibirica*, with small vivid green flowers, in shape like the *Aquilegia ; Aconitum lycoctonum ;* the beautiful spotted *Ophelia bimaculata* from the Yumoto plain ; the small red and yellow *Melampyrum roseum v. japonicum*, and the clover-headed *Poterium officinale*, which covers the plain with great patches of dark-crimson.

.

August 20th.—The Lilies are nearly over, and Nature seems to rest until the Amaryllids are ready. The gap is filled with the scent of the Lotus, the wonderful sacred flower, which comes from afar across the heated plains.

In the crannies of the temple walls, the little red corolla of *Conandron ramondioides*, and everywhere else the giant *Macleya cordata*, which artists have seized on as one of the typical floral features of a Japanese landscape.

August 23rd.—Belated at Shirakawa, and wandering through the country in the fierce heat of the midday sun, I stumble unexpectedly on a bed of Lotus, and I gaze for the first time on the lovely flowers, which, growing in the black mud, are to the Japanese the symbols of purity—" A pure and beautiful woman in a haunt of vice : '—" A man of stainless honour in a wicked world." Such exquisite tender colours, such perfection of form, such stately grace of growth,—set round with mighty and shapely leaves with their under-colouring of pale blue, which seems in the sunlight to reflect the heavens,—has the Lotus, that it is no wonder religion has set it on the highest pinnacle of its symbolism. The beautiful pencilling of the veins on the petals seems to have been the fount of inspiration for the old Buddhist artists, whose work was never perfect until the gold lines on the flower they loved to paint vied with Nature in her accuracy.

In the early morning the rising sun receives a royal salute of welcome from a hundred and one opening buds.

Later on, in the journey to the pine islands of Matsushima, fields of the pink *Amaryllis Belladonna;* and in the country round about, purple *Lobelia sessilifolia* and pink *Lythrum virgatum;* and fields "brocaded with bush-clover," *Lespedeza bicolor.*

August 28th.—Returning to Nikko after my ramble, I find *Lycoris aurea* in full bloom ; and two old friends, the vivid blue Corncockle to remind us of those fields of corn at home, which we miss so in a Japanese landscape, all too green ; and great masses of a *Clematis,* twin-brother to the "Traveller's Joy," *Clematis paniculata,* in whose fragrance, though it be not the true Joy, but rather a more perfect flower in form and scent, home memories are embalmed.

To be noted also, a cluster of the Vetch tribe, *Lespedeza bicolor,* pale yellow and purple, and a white variety : *L. bicolor v. Sieboldi,* crimson and pink : *Vicia cracca v. japonica,* purple and white, trailing in reckless profusion over the hedges : and the buff-tinted *Amphicarpæa Edgeworthii v. japonica.* Also a somewhat scarce white-flowered creeper in the hedges, *Cucubalus bacciferus ;* the large Evening-Primrose, *Œnothera biennis ;* and the minute St. John's wort, *Hypericum erectum.*

LOTUS.

ALFRED EAST

September.

September 14th.—The return to the capital. Last year I remember the country gardens full of Balsams and blood-red giant Coxcombs, and my own a blaze of Marigolds and " Morning-glory," *Asagao* (*Ipomæa hederacea*) ; common enough flowers, but delightful for their starlike multitude, the rains and intense heat of the Japanese summer drawing up seedlings to an almost unrecognizable height. But this year a typhoon of terrific force and long duration swept over the country in the budding-time of autumn flowers, leaving floral death and desolation in its track. The country gardens are bare, and at home, nothing but a litter of dead leaves and broken branches ; *Paullownia* trees uprooted ; the shrubs standing in holes in the ground as if the many-handed spirit of the storm had caught each one and made it dance a pirouette in its bed ; not a flower. " *Et devastavit* " must be written in the indictment of Nature for her wild freak.

Many days spent in clearing away, resetting trees and bushes, and pronouncing sentence against cumbering the ground. One survivor alone I found growing in odd sequestered nooks, the long slender spikes of *Polygonum filiforme*, with its tiniest flowers of crimson, and handsome dark-barred leaves.

September 20th.—Toil rewarded. The extraordinary recuperativeness of the Japanese nature, that delightful power of throwing off the remembrance of troubles past in the joy of present delights, finds its prototype in Nature herself. When the last vestiges of the wreck have been swept away, and all memories of the storm have been obliterated, the garden forgets her sorrow in the joy of decking herself anew. The " Glory of the morning " has renewed its youth, and, sending out fresh

53

growth of stem and leaf, welcomes the sun again with its many-tinted lovely flowers. These my sluggard friend who rises betimes never sees, for they roll themselves up to sleep long before the sun has mounted to mid-heaven. He revels in the " Evening glory," and will not believe the poet when he says, that for all its beauty it cannot compare with the exquisite clouds of colour which deck the path of the morning.

Here, with the poet's leave, I note for its preservation what many another note-book will assuredly have preserved as well :

> Asagao ni
> Tsurube torarete
> Morai midsu.

Which means literally :—

> *Convolvulus*
> *Bucket taking,*
> *I borrow water.*

And as our English poet has rendered it :—

*T*HE " *Morning glory* "
    Her leaves and bells has bound
My bucket-handle round.
I could not break the bands
Of those soft hands.
The bucket and the well I left ;
Lend me some water, for I come bereft.

From Sir Edwin Arnold's "Letters by Sea and Land."

After the *Ipomœa*, the most striking thing in the garden is *Hibiscus coccineus*, with its stalks six feet high and huge crimson
54

flowers; even more beautiful are the delicate white and pink flowers of the low-growing *Hibiscus mutabilis*. *Amaryllis candida*, and a flame-coloured *Ipomœa quamoclit* grown from seeds, a Chinese flower I think, help to keep us gay.

At the festival of the full moon, cakes were made in old days and offered to the Queen of the Night. The floral decorations consisted of seven wild flowers in full bloom at this time : they are the autumn *Nana kusa*. Like the " seven herbs " of winter their names run into a 5, 7, 5, 7, 7 couplet.

> **Hagi, Kikio**
> **Obana, Kusubana,**
> **Asagao no**
> **Hana, Omineshi,**
> **Nadishiko no hana.**

Which being translated botanically runs thus :—

> *Lespedeza bicolor, Platycodon grandiflorum,*
> *Eularia japonica, Pueraria Thunbergiana,*
> *Of Ipomœa hederacea*
> *The flower, Patrinia scabiosæfolia,*
> *And the flower of Dianthus superbus.*

Our Lady's Day.—The fields between Tokyo and Yokohama are resplendent with the crimson Amaryllid, *Lycoris radiata*. Among the Japanese this delightful plant, looking like a huge spider with the long stamens from half-a-dozen flowers ranged in a regular circle round its head, is looked upon as an emblem of death ; it is left to bloom in the fields by the side of the withering Lotus leaves, and is never brought into the house.

55

In the gardens many-hued Coxcombs have revived, and are growing and flowering apace. *Olea fragrans* is full of amber blossom ; and the flaming *Clerodendron squamatum* from China lights up the dark corners of the house. The skies of these early autumn days are serene, the air balmy as if it never knew disturbance ; but the dearth of flowers in the markets, and the giant trees of *Salisburia adiantifolia* all seared on the seaward side, look as if a gust of fire had swept by them, and remind us unpleasantly of those few stormy summer days. The leaves of *Salisburia* are used in the hot weather to keep moth and damp away.

October.

October 25th.—My diary draws to a close. To-day I have a very short list to note : a dark red-spotted *Tricyrtis macropoda*, a clump of double red *Anemone japonica*, and the delicate *Camellia Sasanqua* in several varieties of colour, from white, palest pink, to red. The petals of the red flowers vary in colour considerably, and are often dashed with large spots of white.

November.

> " *Let the Emperor live for ever. May he see the Chrysanthemum Cup go round autumn after autumn for a thousand years !* "

The Chrysanthemums, whose virtue as a specific against malaria when dipped in *saké* is referred to in the above quotation, are in full bloom. They are of every hue and of every

56

shape and size. " The white ones gleam so in the sunshine that
in the early morning I can scarcely distinguish the flowers from
the hoar-frost laid so gently on them." The long thread-like
petals of one class are new to me ; and at one of the shows in
the neighbouring village I have just seen the very latest
triumph of horticultural art, a Chrysanthemum with a small
flower of vivid apple-green. This was its first appearance in
public, and larger flowers are promised in a season or two.
I need not dwell long upon the Chrysanthemum, for the
Imperial emblem of Japan is too old a friend at home to need
special notice, and I am not at all sure that we do not see
a finer exhibition in the Temple Gardens every autumn, so
far as size and colour go, than even in the Palace Gardens
here.

The Chrysanthemum is the last of the " Four Gentlemen,"—
Shi kunshi—so called for the qualities of which they are typical.
The four are : *Mume*, the Plum,—vigour, by reason of its being
the first flower to brave the snows, and sweetness : *Ran*, the
Orchis—grace in adversity, for it preserves all its elegance
though it grows wild on the mountains : *Také*, the Bamboo—
uprightness : and *Kiku*, the Chrysanthemum—the emblem of
To ye mei, a distinguished Chinese official who, many centuries
ago, retired from the Government service on account of its
corruptness.

One or two curiosities I must notice. The Japanese gardener
knows, among his manifold secrets, a special method for pro-
ducing flowers in great numbers on a single stem. In every
garden now there are two or three plants, set in a special place
of honour, each with three to four hundred blooms, not of the
" pom-pom " variety, which is also much grown, but good blossoms
three inches in diameter. There were three plants in the
Imperial Gardens with over four, five, and six hundred flowers
respectively. Another but much less interesting habit is that of
grafting five, six, and as many as ten varieties on one stem.
The result is nil.

I

But the fair in the village of Dana Saka is most curious and interesting. In the booths are represented scenes of Japanese romance ; the figures are either life-size or colossal, the faces and hands of clay exceedingly well modelled, and all the body built up with Chrysanthemums, in many parts growing shrubs being used. My friend Mr. J. B. Rentiers, of the British Legation in Tokyo, has allowed me to reproduce a photograph of the seven Gods of Happiness in their Treasure-ship, which he succeeded in obtaining, in spite of the darkness of the booths.

THE SEVEN GODS OF HAPPINESS, AT OANA SAMA.

Mid-November.—The last pilgrimage of the year; across the paddy fields to Oji, where you come suddenly on an old Temple, and a little secluded valley glowing like a furnace with the colours of the dying Maple leaves. It is the last spot of colour in the year ; in a week or less the earth will have resumed her dun-brown mantle. The cycle of flowers is complete, and my too prolix diary at an end. The Maples gather round the closing scene ; they are the crimson clouds which hang about the sunset of flower-life in Japan.

A CONVENTIONALISED SPRAY OF PAULLOWNIA IMPERIALIS

www.ingramcontent.com/pod-product-compliance
Lightning Source LLC
Chambersburg PA
CBHW021957190326
41519CB00009B/1297